ADDRESS

DELIVERED BY

HON. HENRY H. CRAPO,

Governor of Michigan,

BEFORE THE

CENTRAL MICHIGAN AGRICULTURAL SOCIETY,

AT THEIR

SHEEP-SHEARING EXHIBITION,

HELD

AT THE AGRICULTURAL COLLEGE FARM,

On Thursday, May 24th, 1866.

LANSING:
JOHN A. KERR & CO., STEAM BOOK AND JOB PRINTERS.
1866.

ADDRESS.

Mr. President, and Members of the "Central Mich. Ag'l Society:"

LADIES AND GENTLEMEN: Remote from the theatre of action in the late rebellion, Michigan has experienced comparatively few of the evils that followed immediately in its path. The usual pursuits of peaceful life, were here scarcely disturbed, and by the permission of a Gracious Providence, the industry of the inhabitants of our State was but little diverted from its legitimate channels. Nevertheless, while so many of her patriot sons were engaged in the deadly strife of Southern battle-fields, and the result of the struggle was in the uncertain future, a sombre cloud could not fail to brood over our daily life, interfering with the full enjoyment of the blessings we retained.

Now, however, the roar of cannon and the noise and tumult of war is no longer heard in our land; the scenes of carnage and blood which our once peaceful and happy country has recently witnessed are at an end; the turmoil and strife of armed hosts in deadly conflict have ceased; the public mind is no longer excited, and the hearts of the people are no longer pained, by the fearful news of battles fought, and of the terrible slaughter of kindred and friends. Social order again invites us to renewed efforts in our respective labor and callings: and we are permitted "to beat our swords into plow-shares and our spears into pruning-hooks."

Like the calm and quiet repose of peace when it follows the clamor and din of war, so is the delightful, cheering and invigorating approach of spring, as it succeeds the chilling blasts and pelting storms of dreary winter.

The truth of this is verified to us on the present occasion. We have come together at this delightful spot, and on this beautiful spring day, not only for the enjoyment of a festive season, but also for the improvement of our minds and the increase of our present stock of knowledge on subjects with which our several interests and our respective tastes are more or less identified.

At your request and upon your kind invitation, I am here to contribute my share—small though it be—to the general fund. I should, however, have much preferred the position of a quiet learner to that of an incompetent teacher—to have *listened* rather than to have *spoken.* But being here, it will be my purpose—by your indulgence—to speak, in general terms, upon such topics as seem to me appropriate to the occasion. I shall not presume to theorize, or to speculate; neither shall I travel through unexplored fields with no other guide than imagination; nor shall I attempt to entertain you with any rhetorical flourishes, or figures of speech; but in a simple manner endeavor to give briefly my own views on the several subjects discussed.

The occasion is undoubtedly one affording a wide field for profitable discussion; yet the space which your greatest indulgence can be expected to allow me will render it necessary that I confine myself to a very few topics, and will barely permit a hasty glance at some of those only which may be considered appropriate in this address. You will therefore, I trust, remember that in case I do not refer to subjects which you may deem of importance, it will be from this reason, and not because I may have considered them unimportant.

In the first place, then, permit me a brief reference to this Association, under whose auspices, and by whose directions—acting in connection with the officers of the Agricultural College—this festival is held. Your Society, I understand, extends over the counties of Ingham, Eaton, Clinton, Livingston and Shiawassee, and has been formed for the purpose of combining and concentrating a wider scope of individual action than could otherwise be attained, with a view to an increased interest in the subject of Agriculture and of Agricultural Fairs; thereby recognizing the principle that "in union there is strength."

The effort is not only laudable, but will, I have no doubt, be productive of the most beneficial results. In fact we have in this very effort to bring into notice and give an increased interest to one of our most important branches of husbandry in our State—the growth and production of wool—abundant evidence that such will be the result. By coming together, as on the present occasion, in the spirit of a free, frank and social interchange of ideas, an increased interest cannot fail of being awakened, as well as an extensive inquiry instituted, among farmers generally, not only as to the most desirable breed of sheep, but also as to the best modes of tending and keeping and feeding the different kinds, with a view to the greatest profits. The influence of such a gathering as this is of much value—not only in encouraging a desire for excellence and creating a spirit of competition and of laudable emulation, but as furnishing the means for an active exchange of the more desirable specimens. Those who assemble are enabled to enjoy a season not merely of relaxation from toil, but also for mutual consultation and discussion; and a healthy and growing interest in everything pertaining to Agriculture, in all its varied forms and branches, is thereby induced.

In this connection I may be permitted to make a few remarks in relation to the salutary influence which our Agricultural Societies cannot fail to exert upon the farmers of Michigan, and of the *benefits* which are certain to flow from them.

There is no employment which keeps man so isolated as that of Agriculture; and these societies serve, in a very great degree, to counteract the bad effects of this by bringing mind into intercourse with mind. They should receive the united and cordial support of every farmer.

Whilst professional men are brought into frequent contact with each other—and the trader is in constant intercourse with his customers—and the mechanic is associated with those employed with him in the shops—the farmer spends most of his time with his family, and with his flocks and herds, and sees comparatively little of others. The Agricultural Fair brings—or should bring—all the farmers together, with their wives and daughters, where a healthy, social intercourse is enjoyed. There a higher standard of excellence in everything is formed. He there learns that what of his own he had been led to believe was the best—whether in flocks or herds, or farm products—may be greatly improved, and his ambition and pride, as well as his interest, are at once excited to make an advance. At the same time the industrious housewife, and the blushing Miss, by an examination of the cloths and flannels—the carpets and quilts—the embroidered skirts and capes—the collars and slippers, discover that these articles are worthy not only of their admiration but of their emulation, and they, too, resolve to copy from a standard of merit higher than their own. Thus is excited among those so brought together a spirit of competition, and a desire in their turn to excel.

Another important benefit resulting from Agricultural Fairs, is a more rapid and general diffusion of knowledge among the farmers in regard to the advantages and practical utility of new inventions, for the saving of time and labor in agricultural operations. This is illustrated very clearly by the exhibition of Mr. Parish's "Stump and Grub Extractor," on exhibition here. This machine, I understand, was patented on the first day of the present month, and *now* all in attendance at this Fair have had an opportunity of witnessing its operations and judging for themselves of its merits. An effective machine of this kind is of incalculable value to the farmer in removing *at once* from his fields the unsightly stumps that disfigure them, and which adds so much to the labor of cultivating those fields. Of the machine itself, I may be permitted to say, by way of digression, that it surpasses in the effectiveness of its operations anything of the kind which I have yet had an opportunity of witnessing.

But this is not all. The mutual consultation and discussion consequent upon Agricultural Fairs, begets a spirit of inquiry and a desire for information in relation to every subject connected with the farmer's calling, and to gratify which he has recourse to periodicals and other works in which its various branches are discussed and explained. He will there learn what agricultural chemistry has done for him, and the importance and value of the analysis of the different kinds of soil. He will also find the result of the various systems of husbandry practiced by others as well as the effects of experiments made, and thereby secure to himself their benefits without incurring their cost. And although no amount of reading alone can make a man a farmer, yet the knowledge derived from a perusal of agricultural papers devoted to the interests of the tillers of the soil will be of incalculable value to him.

SHEEP-HUSBANDRY.

It will undoubtedly be expected that "Sheep-Husbandry," not only from the importance of the subject itself, but because of its being the principal feature in this exhibition, should receive at my hands a due share of consideration.

I am free to confess, however, that the subject will be approached with no small degree of hesitancy and distrust on my part, not only because of my want of practical knowledge in regard to it, but also because it may be fairly regarded, I think, in many respects at least, as a sort of debatable ground.

Different views are undoubtedly entertained by equally intelligent and experienced men, upon this as well as upon other equally important subjects; and the fact I believe is well established that "Doctors" not only *may* but *do* very often "disagree," and that, too, sometimes very tenaciously. Should I advance opinions at variance with those entertained by well-informed and practical men who may listen to me, I will simply remark that I am not here to lay down rules and establish principles for the guidance of any one, but to discuss principles and rules of action, as well as practical questions, with a view to lead others the more carefully to inquire into and investigate the same.

The subject of Sheep-Husbandry with us is certainly an important one—wool being a great, leading staple product of our State; and very much attention is now being paid to it, which is fully justified by the advantages of our soil and climate for the keeping of sheep. The farmers of Michigan are fully aroused to the importance of this interest, and have labored zealously, and at much expense and cost, to improve their breeds of sheep, and to foster and develop this great interest. They have certainly done much in this direction; but more—very much more, I apprehend—remains yet to be done.

It must, however, be remembered that a blind zeal, without that knowledge which is the result of experience, observation and study, will do very little in the right direction.

Sheep, like cattle, should be selected for specific purposes, and in reference to affording the greatest profit under existing, and probable future circumstances. The exclusive cultivation of this or that breed—of the fine or coarse, or of the long or short wools—whether kept exclusively for their wool, or both for their wool and the shambles, should never be practiced, unless under special and unusual circumstances. The farmer in this, as in every other agricultural department, must endeavor to see his relation to the merchant, and adopt a practice having in view the chances of ultimately reaching the most certain as well as the most profitable market; for, after all, the connection between the producer and the manufacturer and merchant, is but a partnership for loss and gain. The merchant will call upon the manufacturer for such woolen goods as his market demands, irrespective of the mere opinion which any one may entertain in favor of this or that kind of wool; and the manufacturer, in his turn, will call upon the farmer for just what is wanted. The farmer should therefore, in the selection of his flocks, have in view the market upon which he is to rely for the sale of his wool; the texture and weight of fleece; the health and vigor of body and constitution, as well as the habits and economy of the animal. He should sedulously seek to bring his sheep to a high degree of perfection in every respect. In seeking to obtain quality of fleece it is a self-evident fact that he should not overlook quantity; and that quantity should also be considered in connection with quality.

It is a patent fact, of which if we needed evidence it may be found in this exhibition as well as in the numerous county exhibitions of similar character, which have recently been held, where very rarely any other class of sheep are seen, that a strong preference for fine-wooled Merinos is very generally, if not almost exclusively, entertained at the present time among the farmers of this State, and money in the purchase of that class is of but little account. It is well known that very high prices are being paid not only for single specimens but for whole flocks of this breed. This is probably all right, so far as it is necessary for the purpose of attaining excellence in flocks, upon points already spoken of. To such a preference there should be no objection, if it be not carried so far as to superinduce an unprofitable reaction—and provided that the demand for the grade of wool produced by these sheep is to have no limit, and that all which can be grown is sure always to command a remunerative price. But will this probably be so? Let us consider.

As I have already intimated, the demand for any particular quality or kind of wool will not depend upon the fact that farmer A or farmer B has such wool to sell, taken from sheep for which he paid very large prices, and of which he has now a very large flock; but rather because that particular kind and quality of wool is called for by the manufacturer simply to fill the orders of the merchant, who in his turn is only desirous to supply the demands of the consumer.

From an examination of our imports, it appears that in 1863, of *sixty millions* of woolen goods, about *forty millions* were manufactured of the longer worsted wool. This wool is required to make a fabric of lustrous appearance for imitations of Alapaca, and for a supply of which our manufacturers now depend mostly on foreign countries. The price of combing wool has been for some time increasing rapidly, in comparison with other wool, in consequence of its consumption gaining upon its growth. And I saw recently that the British farmer had been urged to increase the production of this article to its fullest extent, both from a consideration of duty as well as of interest.

The manufacturer of Alapaca cloths—a most beautiful fabric of recent introduction—and their extensive use, has not only led to this increased demand, but has enhanced the price of this kind of wool, which will undoubtedly be maintained, as new fabrics requiring to be made from long wools, especially for the garments of ladies, are now being introduced in great variety, and are becoming daily more popular and of more general use. Another cause for the continued and increasing demand for these wools is the facility with which they can be used for the purpose of making imitations of Lama fabrics and Alapacas; and I have no doubt that factories for the manufacture of these goods will rapidly multiply in New England and elsewhere, and will soon, to a very great extent take the place of those now consuming the fine wools.

In support of these views, permit me to give the following extracts from the work of Mr. Randall, the well known and enthusiastic champion of the Merinos. He says:

"In the American market there is a much larger demand for medium than fine wools, and the former commands much the best price in proportion to its cost of production."

Again he says:

"American producers of very fine wool have ever fed an expectation, but never obtained the fruition of their hopes."

These are significant admissions, coming as they do from such a quarter.

The South Downs are a variety of sheep of decided merit; but have never, I think, been fully appreciated by the farmers of Michigan. They are of large size and symmetrically formed, with hardy and robust constitutions, and their wool is fine, short and curled, and destitute of fibrous spires that give to it the felting properties. It is neither a short nor a long staple, but ranks in this country as "middle wool." The shorter staples are made into flannels and light woolen goods; and the longer are extensively used for combing. Their mutton is unsurpassed; its flavor is delicate, and the flesh juicy and well intermixed with fat. They are the most prolific breeders—the proportion of ewes bringing twins being at least fifty per cent. I recently saw a fine flock of South Down ewes in the State of New York of which more than three-fourths of the whole flock had twins.

Among the more desirable varieties or families, for the production of long wool, in this climate, are, perhaps, the Cotswolds, noble specimens of which you have had an opportunity of inspecting on this occasion; and have, I trust, with me, been highly gratified at their weight of carcass, combined with their fine forms and apparent hardiness of constitution, as well as the superior fleeces they have now yielded.

My purpose, however, is not to advocate the claims of this or that class of sheep at the expense of any other, but to present such views for your consideration as may lead to a more thorough and candid investigation of the whole matter.

Let me say in continuation of this subject, that in a comparison between the Cotswold and other long wool varieties, with the fine wool Merinos the *advantage as to weight of fleece* is decidedly with the former; and especially so when their respective fleeces are thoroughly cleansed and scoured; for whilst the loss of the long wools very rarely reaches *twenty per cent.*, that of the Merinos generally much exceed *fifty per cent.*, and the fleeces of prize rams often more than *seventy per cent.* Manufacturers are already beginning to make a discrimination between wool that is clean and that which is not so. Suppose they buy the South Down, Cotswold and Leicester wools, and their grades, from which is lost by scouring twenty per cent. only, whilst upon the finest Michigan wool there is lost *fifty* per cent. and more—making the cost of the latter, at ordinary prices, one-third more per pound than the former, how long will it be before they will study to increase their consumption of long wool when they can make from *thirty* to *forty* per cent. more cloth with the same money? They will certainly seek to avoid, in some way, the necessity of buying with their wool so very large a per centage of grease and dirt, as they claim they are now doing in the purchase of fine wools.

The South Downs, as I have already stated, as well as the long wool sheep, have a decided advantage in the quantity and value of meat which they yield for the shambles; for no one, I apprehend, will deny the fact they not only yield more wool but very much more flesh to the live weight than do the Merinos. And this is a fact worthy the serious consideration of farmers, and certainly a strong argument in favor of the more general breeding of long wool sheep. The war, and perhaps other causes, have very seriously reduced our supply of meats, the waste of which cannot soon be repaired. Many of our soldiers will not again return to rural life, which will be quite too tame for them after the long, protracted excitement of war. They will seek other occupations, and be consumers rather than producers of meats. In addition to this a tide of foreign immigration is setting in upon our shores, where they will continue to swarm for years to come as never before, hungry for meat; and it has been conclusively demonstrated that the ratio of our ordinary increase of population far exceeds the production of cattle and sheep, which deficiency in beef and mutton must hereafter be supplied in some way. I will again quote from Mr. Randall's work. He says:

"I am strongly impressed with the opinion that the production of mutton has been too much disregarded as a concomitant of the production of wool. Near large meat markets mutton is the *prime consideration*, and wool but the accessory."

Here, then, is a potent combination of circumstances, which were never before brought together, guaranteeing an abundant remuneration, as I believe, to those who may engage in this particular branch of husbandry; and the field, although now new, will nevertheless, I have little doubt, be very soon successfully occupied. I cannot but hope that our ambitious and enterprising stock breeders will secure to themselves their full share.

Perhaps I have already exhausted your patience by dwelling so long upon this subject; but regarding it—as I most certainly do—as a very important one, and this being an appropriate occasion for its discussion, you will, I trust, bear with me a moment longer, whilst I venture to make a few practical suggestions, before taking leave of it. Let me then say, in this matter of Sheep Husbandry, in addition to what has already been said, that you should guard against extreme views of any kind. Merinos are undoubtedly a valuable and a very desirable breed of sheep, as witness the noble specimens exhibited on this occasion; but you do not want them and nothing else, unless they will pay a better profit than any other sheep; nor should you pay an extravagantly high price for them merely to enrich the sheep-breeders of another State; nor because it is fashionable to do so. You should remember that the South Downs, the Leicesters, the Cotswolds, as well as some others perhaps, also have their respective claims to favor and are worthy of your consideration. My own opinion is that a grade of sheep may be produced by a cross between the Costwolds and some other varieties, which will furnish a staple of fine, long, combing wool of lustrous appearance, that will prove—all things considered—quite as remunerative as fleeces from the choicest Merinos and their grades.

You should, also, avoid the too common error of overstocking with sheep when the price of wool is high. Sheep Husbandry has been a very profitable branch of business for the farmers of this State; but like every other business it may be overdone, and is liable to fluctuations and changes. Sheep must be well fed and cared for in order to produce heavy fleeces; and there is certainly a limit to the number which may profitably be kept upon any farm; and it not unfrequently happens that a flock of fifty sheep on a small farm, will yield a larger net profit than would a flock of five hundred if kept upon the same farm.

When the price of wool is high, the farmers are too reluctant to sell off their sheep, and thus become liable to an overstock. In fact, this is now the great danger of the wool-growers of Michigan. The best economy, and the most judicious management, will be to keep down the number of your flocks to your means of pasturage and feed; and constantly aim to improve the grade and quality of those you retain by disposing of the less desirable specimens for mutton. Your motto should be to elevate the standard of your flocks, rather than to increase their number beyond your means of feeding.

Another evil is also to be guarded against,—that of giving your attention to sheep to the exclusion of cattle. I am aware that in the past there have been—in this State—few advocates for the raising of cattle, and that the sound judgment of any man would at once be brought into question who should attempt to do so. But I think there has been more of prejudice than reason in this. The farmer, as a mere matter of policy, should not confine himself to any one thing, as thereby the fluctuations and changes incident to any branch of business, may very possibly—nay very probably—disappoint his hopes and expectations. If he has only sheep on which to rely, a sudden fall in the price of sheep and wool, or a general prevalence of any of the diseases to which sheep are always liable, would be a serious disaster to him; whereas, if his attention is directed to both sheep and cattle, as well as to horses, swine, &c., his chances of certain and continued success are very greatly multiplied. In fact, cattle are already commanding enormous prices in consequence of a general scarcity everywhere, not only for the shambles, but for the dairy, and this deficiency will not, I apprehend, be very soon supplied. I have recently visted some of the more highly cultivated portions of the State of New York, where I found good fair cows were worth *one hundred dollars* each and not easily to be had at that. Good sized, first quality working oxen, are now worth here $250 per pair; and a large lot of cattle has recently been sold for beef in Flint, at seven cents per pound, live weight. Horses, too, are scarce, and must continue to be so for a long time, as their destruction by the late war was very great, and years will be required to replace those so destroyed, especially in the rebel and border States, which must be supplied from the North. Swine, also, are now deficient, and principally because, a few years since, for a time the price of pork was very low, and their growth was in consequence, at once almost abandoned. The farmer should take a broader view of things, and pursue a steady, onward course, avoiding all extremes, as well as sudden changes. As a large portion of his farm products are more adapted to the feed of cattle, horses and swine than to sheep, he should, if for no other reason, keep a due proportion of these animals, any excitement in favor of sheep notwithstanding. My own opinion most decidedly is, that the time has come when the best interests of the farmers of Michigan require that a portion of the attention now being devoted to sheep husbandry should be directed to that of other kinds of stock.

But, to return again from this digression to the subject of sheep and wool.

One of the most serious difficulties with which the farmers have to contend, is the combinations that are too often sought to be made by purchasers to secure their wool at the lowest possible figures. The manufacturers and wool buyers, undoubtedly act in concert,—at least to some considerable extent,—to depress the price, and especially so, before and about the time the new clip is coming in. They are well drilled in this, and many of their operations are systematic and efficient. At such time they pretend not to be in want of wool,—that the demand will be light, &c. Purchases are made very sparingly, and temporary supplies are procured from other sources, even at a higher cost than the farmers ask. This is done upon the ground that an occasional sacrifice of this kind pays well in the end, if thereby they are able to keep down the price of the great bulk of domestic wool. Sometimes fictitious sales are reported, and various other means are employed to this end, with the view that a few holders, at least,—either from necessity or timidity,—may be induced to *sell*, and thus aid their efforts to establish low prices.

It thus becomes the duty of the farmers to act with much consideration, study and wisdom; and purely as a matter of self-defense, to adopt some concert of action among themselves for the protection of their own interests. When the price is low and the market dull at the time of shearing, there should not be too much haste in making sales. In 1861, I think it was, the farmers were over anxious to sell, for no other reason than because at that time the price of wool was very low and the market dull. They then overlooked the well established commercial fact that depressed markets generally advance, rather than retrograde, and that Government disbursements then certain to be made would create funds and a higher market, and that the demand for the staple would increase. They consequently sold for *twenty-five cents* per pound, fleeces, that in less than three months commanded *forty-five* to *fifty* cents. They also, in many instances, offered to sell their fleeces for less than half the sum they would bring in a very few weeks. On the other hand, as is too common, when wool at the time of shearing commands a high price and the market is brisk, the farmers are inclined to hold on for still higher prices. But this is another mistake in the opposite direction. The rule should be,—"*sell*" when the market is quick, and prices are good;—and "*hold on*" when the market is dull and prices are low.

Before leaving this subject, permit me to call your attention to another important matter in connection with sheep husbandry in this State. Sufficient care has not heretofore been taken to clean and otherwise properly prepare this great and important staple for market, and the consequence has been that the character and representation of "Michigan wool," I am sorry to discover, has been very seriously lowered in the market, and a great loss to the producers has thereby been sustained. It is a fact, perhaps not generally kown, that from this cause alone, "Ohio wool" sells for about *five* to *ten cents* per pound more than "Michigan wool." In an interview which I recently had with an extensive eastern manufacturer, who was induced last season for the first time to purchase a lot of "Michigan wool," he expressed his surprise that the Michigan wool growers should be so heedless of their own interests as to overlook this important fact. From his statements I learned that the prejudice of the manufacturers against "Michigan wool" was so great that many of them would not buy it at hardly any price when they could get "Ohio wool." He said a large proportion of our wool was poorly washed, and that this was true of a great proportion of our finest and best lots; and that it was not only sent to market in this condition, but was badly and slovenly put up, with much larger twine than they use in Ohio,—the fleeces, also having a torn and jagged appearance; and many of them, when opened, were found to contain the *unwashed* tags. He, however, expressed himself highly pleased with the quality of the wool he had purchased, and said it compared favorably in that respect with any he had ever received from Ohio; and he believed if our wool could be sent to market as clean and in as good condition otherwise as the Ohio wool,—and the prejudice which has been created against it, in consequence of this not having been the case heretofore, could once be removed, he doubted not that "Michigan wool" would command in the market the highest prices and the most ready sales.

This is certainly a serious matter, and prompt and efficient measures, of some kind, should at once be taken to remedy the evil; and every wool-grower should feel, as he really is, personally interested in the work. I commend this subject, gentlemen, to your serious consideration, and trust some concert of action will be had to prevent a continuance of this great evil, and to place "Michigan wool" where it should most certainly stand, at the head of the list. If this can be done in no other way, I would suggest the formation of a "Wool-Growers Board of Trade," or some other efficient organization for the purpose—if for no other—of tracing out and holding up to scorn every individual who shall aid in inflicting so serious an injury to this great interest, and of doing so great a wrong to his neighbors and fellow-citizens, and that, too, from the base and fraudulent motive of selling dirt and tags as fine wool—for be assured that any imposition of this sort, practiced upon manufacturers, will recoil upon our own heads; and where *one* cent will thus be saved, thousands, yes, tens of thousands of dollars, will, as a necessary consequence, be indirectly lost to the farmers of Michigan. And the loss they have sustained from this cause during the last three or four years will undoubtedly exceed the enormous sum of two millions of dollars.

But I must take leave of this subject.

THE STATE AGRICULTURAL COLLEGE.

Permit me now to occupy your attention for a brief space whilst I speak of this Institution—the State Agricultural College—upon whose grounds we are now assembled, and where by the kindness and courtesy of its officers, we have been so cordially welcomed and so pleasantly entertained. It is not, I think, inappropriate to the occasion that I should do so.

Let me remind you then, in the outset of my remarks on this subject, that this Institution is in its early infancy; and that notwithstanding the beautiful landscape which is spread out before us; with its verdant fields just springing into luxuriance, dotted with the finest specimens of the choicest breeds of sheep and cattle, with the College grounds skillfully laid out and now in process of being tastefully adorned by Art, a few years only have been numbered with the past since not only this spot, but all the surrounding country, as well as almost the entire territory of our young, but noble and now highly prosperous State, was an unbroken wilderness, covered with the primeval forest, the entangled woods giving shelter and concealment to wild and ferocious beasts, as well as to the wandering and savage red man. What a change has thus been wrought in a few short years! the result of the toil and privation of the adventurous pioneers, of whom many have already become intelligent, enterprising and forehanded farmers.

And more than this: Michigan, although but recently settled, and one of the youngest in the great sisterhood of States, has been the first to establish a professional school for the agricultural education of her sons, in which is not only taught the sciences and their application to agriculture, but also agriculture as an art, with such experiments as are calculated to impart a more thorough and practical knowledge of the same; and connected with the study of these a department of manual labor; the legitimate effect of all which is to increase the student's desire for knowledge as well as his love of study, and to remove the barrier too often existing between the educated and laboring classes—which can only be done by giving a better education to those who labor, and by removing the prejudices of the educated against labor.

But I propose to speak more definitely of the aims and objects of this Institution, as well as its claims to the favor and support of the farmers of Michigan. They need not be told, I think, that its design is to promote their benefit. But have the farmers of this State, as a class, heretofore recognized this fact? And have they in return for the advantages which it proposes to them, given it that countenance and encouragement which it claims at their hands? I fear not. There are, it is true, noble exceptions to this; yet it is also true that a large proportion of their number have looked upon it with suspicion and distrust, as though its purpose was to do them a wrong—to inflict upon them an evil. They have not merely withheld from it their aid and support, but their active influence has too often been exerted to its disadvantage and prejudice. This is certainly wrong—very wrong!

Let us look a little into this matter. Is knowledge—a knowledge of those sciences which are intimately connected with agriculture as an art—of no value to the farmer? Is it necessary that he should be a dolt in order to be fitted for his vocation? Will ignorance and bad husbandry increase his crops or enable him to find a better market for his products? Or, will his enjoyment, in his daily round of toil, be any greater because unconscious that he is groping his way along in the dark? No! For however that may have been in the past it is certainly not the case now. And although "ignorance," as it is said, may be "bliss," yet in these days, at least, it must be a sort of negative bliss.

Ignorance is certainly not power; nor does it lead to wealth as a means of comfortable support and enjoyment—which is the legitimate end of all labor. Will *ignorance* give respectability, or sweeten the toil of the husbandman? Will it elevate his thoughts and desires to higher and nobler aims, or inspire him to "look from nature up to nature's God?" Will it lead him instead of a fixed stolid gaze upon the earth over which he walks, to engage in the study of those great and omnipotent laws which regulate all matter, and which so wonderfully, yet certainly control both the animal and vegetable kingdoms? No! It will accomplish none of these desirable ends, but the very reverse of them all.

This proposition is so self-evident to intelligent men, that to advance it to such an audience as the one before me—except as the basis of an argument—must be entirely superfluous. But what was the social position of the farmers, let me ask—even in this highly favored country—fifty or sixty years ago? Were they not then regarded as men without knowledge—devoid almost of sensibilities—unfitted for anything except the mere routine of daily labor and toil—and capable only of delving in the soil day by day? And were they not then considered, even by themselves as well as by others, as occupying the very lowest position in the scale of society? Such were the facts. Every person who was regarded as too ignorant and uncultivated for other pursuits, was, by common consent, considered as having a prescriptive right to farming as a vocation. In fact ignorance was regarded as the proper and sufficient diploma for the farmer. And as a consequence he was not only poor and without influence, but too often considered by others as without respectability merely because he was a farmer; and all that was conceded to him—in fact all that he claimed for himself—was a simple subsistence upon the hardest fare, without any of the luxuries, and very often with a scarcity of the necessaries of life.

Remember, I am speaking of the farmers, as a class, *fifty* or *sixty* years ago—before there were any county fairs, or agricultural colleges, newspapers or magazines, and when agriculture was the result of labor without knowledge, system or calculation.

But although the farmers have emerged from this condition very slowly, yet what is their position now? Are they not regarded as being on a level at least with those of other callings in social importance? Do they not occupy positions of confidence and trust in society? Are they not found in our Legislative Halls in fair proportion with men of different pursuits? This is certainly true: and the advance alone is the result of a higher mental culture—of a wider range of thought—and of an increased fund of knowledge, and consequently of an improved system of farming.

And if the advance of agriculture and the condition of the farmer have been tardy, as compared with the improvement in other departments of labor—in other avocations of life—it is solely because science and study have not as soon been applied to agriculture—and because also the farmer has not been permitted the advantages resulting from so early a development of facts connected with his calling as have other classes of men.

But the great work is now fairly in progress of elevating the farmer to his true position in the social order of society—of teaching him that his vocation, instead of being the dull, unintellectual lot of the ignorant, is the most noble and dignified, as well as the most conducive to men's happiness in which he can be engaged; and nothing is now wanting to secure the steady advancement of this work, but for the farmers to do justice to themselves and to their calling, by laying hold of the means for that end which are placed within their reach. Assuming all this to be undeniably true, where can be found more potent agencies in the work of elevation than Agricultural Colleges? And why, then, should any farmer in this State hold back from giving this Institution his cordial and hearty support? And stranger still—why should he put himself in antagonism to its success? Such an attitude, to my mind, is not merely unwise, but preposterous—yes, suicidal. If the College is not what it should be, the more his self-interest should prompt him to bestow upon it his aid. It is the *Farmers' Institution*—founded for *his* benefit, at much cost; and if *he* does not feel an interest in it and labor to make it a success, who will? Who should?

But why have a portion of the farmers of Michigan seemed to look with distrust upon this Institution, and in some cases, I regret to say, seemed to regard it as a sort of wrong to themselves; and if they have not actually opposed, have, at least, withheld from it their support? I must confess, that should I give what seemed to me to be the true answer to these questions, it might be regarded by some who have not very carefully looked into the subject, as an assumption on my part unwarranted by facts.

Would that it were so; that I were mistaken. But having given the subject some little thought and investigation, you will, I trust, permit me the honest expression of my own views upon this important matter. It is for that purpose and none other, that I am here. But you, Mr. President, as well as all those now present, can certainly take no personal exception to these views, as the very fact of such presence shows that you are not of the class to which I may allude; and I am gratified in being able to say that I believe there are very many others, not present, who are the warm and devoted friends of this Institution; and who, with you, I most certainly hope, constitute the rule and not the exception.

But the answer: And in giving which, I will avail myself of the privilege conceded to a certain class of men,—that of answering one question by asking another. Why then do men ever oppose or neglect their own interests? To my mind, only from want of knowledge, from prejudice or self-will—or some other of the same brood of enemies to man's success in laudable undertakings; and of which *ignorance* is the chief, and may be regarded as the prolific source of all the others. In this case, undoubtedly, as in others, some are opposed from a mere notion of opposition, or from a mere whim; others again, simply to agree with, or differ from, some, who are either in favor or opposed; whilst some must oppose whatever they themselves do not originate;—and, others again, have no doubt been led honestly to entertain a distrust which has finally grown into an opposition, through the influence of misrepresentations, or from a perversion of facts by those whose interests, from some cause, are at variance with its success.

But I am quite certain that the whole opposition and indifference to this Institution, so far as it may come from the farmers themselves, is unnatural and fictitious, and will soon pass away as does everything else which is built upon such foundation. It is said by some that "the Institution has been a mistake from the beginning;" that it "was located wrong;" that it "was not started right;" that it "has been badly managed;" and that it "is an expensive concern, and will never pay;" and a great deal more. But it is very easy to say all this, and yet there may be very little reality in it, and still less reason.

Let me here say to the objectors and fault-finders,—suppose all this be true? who *then* is to blame? Is the Institution itself responsible for all these mistakes? Or, are they not rather the consequences of unavoidable and untoward circumstances, magnified and aggravated by *your* opposition, and over which its friends and managers could have no possible control. I admit the probability that the early success of the College would have been more certainly secured, had an old and highly cultivated farm been purchased for the purpose; but for this the means were wanting. You say, perhaps, that College students should not be required to *clear land and dig stumps*. True; but when the officers and managers of such an Institution are *compelled* to do this, and to reach the end desired as best they may through such means, they are certainly entitled to all praise, and richly deserve the meed of commendation for even partial success, and which should be all the dearer to us because of being reached under such adverse circumstances. That the facilities which the College now possesses are inadequate to the proper accommodation of those who wish to avail themselves of its advantages, and even to the extent of the limited number of students now belonging to it, is certainly to be regretted. But this is an evil to be overcome by the patient and persistent efforts of its friends, and not by the antagonism and opposition of its enemies; by making the most out of the limited means at command, and not by abandoning the whole because the means are not now all we could desire. That its management may have been a matter of criticism with those who have known but little about it, or who have taken little or no pains to investigate the facts, is not strange; yet, for one, I am clearly of the opinion that—when all the difficulties with which it has had to contend, are duly considered—its management, thus far, has been all that any person could reasonably hope for or expect; and more—that its officers and professors are entitled to great credit and much praise, for securing under so much discouragement, that degree of success which is apparent here even to the casual observer; and claim of us, and are entitled to receive at our hands, a proper and just recognition of their valuable services, and the fidelity with which they have been rendered.

Farmers of Michigan! Be not led astray by such objections as I have stated, or by any others of a similar import. You have here a noble Institution, in faithful and competent hands—one that will soon be of incalculable value to you—and one, too, that will reflect much credit not only upon you, but upon the whole State. And although it may not now be all you could wish or desire, yet when we consider what it now is in view of the difficulties with which it has had to contend, we have a sure guarantee, that it will yet be a success and will realize all your reasonable expectations. Let me ask of you, in all earnestness and candor, to give it *now* your warm, your hearty support, so that you may not only assist in securing for yourselves and the public the great end of its establishment, but that you may, by and by, safely, and without the fear of successful contradiction, lay claim to the honor of being among its early friends and upholders. There is something noble and magnanimous in rendering substantial aid and support to a cause in the hour of its weakness and in the time of its need; whilst there is something not only selfish but mean, in stepping forward with proffers of assistance, and with spurious claims of imaginary or intended favor, when such assistance is no longer needed, and when the heat and burden of the day has been borne by others; for, be assured, that the time is coming when no farmer will covet the distinction of having been among the number of the enemies of this Institution.

The advantages of our Agricultural College, in connection with an experimental farm, are too obvious to every intelligent mind to require that I should occupy your time in dwelling upon them. And, when I speak of an experimental farm, I do not mean a mere model farm, by which a specimen of good farming only is exhibited; but, like this, a farm embracing a variety of soils—adapted to an extensive range of experiments—and where the value of the different kinds of grain may be tested, as well as the relative advantages of different modes of tillage; the relative effect and value, by actual trial, as well as by analysis, of various manures as fertilizers; and the economy of labor; as well as the comparative value of the different breeds of cattle, sheep, horses, swine, &c., &c., with a view to the introduction and dissemination among the farmers of the State, of such as should prove the most profitable; or of such as could be most successfully used for obtaining the most desirable grades. Such a farm as this, under the efficient and skillful management of its present able and persevering Superintendent, cannot fail to be of very great benefit to the farmers of this State, and should, both as a matter of duty to others and of interest to themselves, receive their united and generous support. And I am firmly of the opinion that when they shall afford this Institution such aid, it will soon become one of the first among our noble institutions of learning, and will be a just cause of pride, not merely to the farmers themselves, but to every intelligent person throughout the whole extent of our noble State.

And now let me invoke, for the future prosperity and success of this College, not merely the liberality of the farmers—or what they may regard as such—in the payment of a trifling tax for its maintenance, but what is of equal importance, and which it has a right to demand in justice to itself—their earnest advocacy of its claims.

AGRICULTURAL COLLEGE STUDENTS.

But I have already, I fear, trespassed quite too far upon your patience, and should, perhaps, before this, have relieved you from further infliction. Yet seeing before me, many—if not all—of the students of this College, I must beg your indulgence for a moment longer, whilst I address to them a very few remarks.

Let me say, then, to you, young gentlemen, that you are now in the enjoyment of privileges for the acquisition of that knowlege so essential to success in after life, which were denied to me—and the absence of which I have felt as a great and serious loss through the whole period of my existence. See to it that you place a just value upon these privileges, and that you do not abuse them. Whilst most of you, I trust, are fitting yourselves for the employment of farming as an avocation, some, perhaps, may be looking forward to other professions and pursuits. I, however, on this occasion, must confine my remarks to those of the former class.

And to such I would briefly remark, that the value and importance of an agricultural education to the youth whose lives are to be devoted to the highly reputable occupation of farming, begin now to be admitted, and happy will it be for our common country, when such education shall be regarded as a necessity. Labor is no longer degrading, but is creditable and dignified; and agricultural pursuits are no longer regarded as disgraceful or ignoble by any except the fop and the coxcomb, but are of all employments the most honorable in which men can be engaged. Nor is it, as has been too often supposed, a cheerless life of toil and fatigue, but has many substantial and endearing charms. It is also the fountain-head for the supply of all our wants; and when contrasted with other employments, its advantages cannot fail to be appreciated. Whilst those who seek a profession must be content to spend many weary years of wasting study—of constant struggle—before they can begin to live, the farmer has at once before him, health and quiet, ease and contentment, as well as the enjoyment of sober pleasures which do not cloy, and whilst the chances of those who engage in commercial pursuits are, that about *ninety-five* out of every *one hundred* are destined to failure, the farmer is exempt from such a hazard, for the chances of failure with him are found to be only about *four* in every *one hundred.*

I do not, of course, in this comparison, include those who, having no land of their own, are obliged to toil for others as laborers, and who cannot therefore be ranked as farmers.

To the farmer, if each day does bring its labors, it also brings its pleasures; and even as he toils in his dusty fields, he can derive unalloyed pleasures, not only from the study and care of his bleating flocks and lowing herds, but from the prospect of an abundant harvest as he looks over his fields of waving grain or contemplates his orchards of rich and luscious fruits. And each day renews to him these pure and substantial pleasures, which afford not only gratification, but health. With the farmer there are no all-absorbing cares, no corroding anxieties, no vitiating excitement. He is measurably freed from the seductions of enervating pleasures. From the green fields and fresh air he drinks constant draughts of inspiration. His great study is, or should be, Nature and Nature's God. To him each season has its profits and its pleasures; for he knows that while he rests or sleeps his fields are working for him. He is also freed, in a great measure, from the baleful influences which attend that false ambition so often excited by other pursuits.

My young friends, when you leave your "Alma Mater" and fix upon your route for life's journey, let your choice of a profession be carefully and wisely made; and then, with undeviating course, pursue it steadily and persistently to the end, for in this only will be found your reasonable chances of ultimate success.

Mr. President, I have already detained you and this audience quite too long; and with many thanks for your kind and patient attention, I will now bring my remarks to a close.

www.ingramcontent.com/pod-product-compliance
Lightning Source LLC
LaVergne TN
LVHW020634110826
845149LV00004B/1188

* 9 7 8 1 4 1 8 1 9 1 2 8 3 *